SCIENTIFIC AMERICAN | EDUCATIONAL PUBLISHING

SCIENTIFIC AMERICAN INVESTIGATES THINGS THAT LIGHT UP THE SKY

LIGHTNING

BY JENNIFER LOMBARDO

Published in 2025 by The Rosen Publishing Group
in association with Scientific American Educational Publishing
2544 Clinton Street, Buffalo NY 14224

Library of Congress Cataloging-in-Publication Data
Names: Lombardo, Jennifer, author.
Title: Lightning / Jennifer Lombardo.
Description: Buffalo, NY : Scientific American Educational Publishing, an imprint of Rosen Publishing, 2025. | Series: Scientific American investigates things that light up the sky | Includes bibliographical references and index. | Audience: Ages 8. | Audience: Grades 2-3.
Identifiers: LCCN 2024027728 | ISBN 9781725352179 (library binding) | ISBN 9781725352162 (paperback) | ISBN 9781725352186 (ebook)
Subjects: LCSH: Lightning–Juvenile literature.
Classification: LCC QC966.5 .L66 2025 | DDC 551.56/32–dc23/eng/20240626
LC record available at https://lccn.loc.gov/2024027728

Portions of this work were originally authored by Kristen Rajczak and published as *Lightning*. All new material in this edition is authored by Jennifer Lombardo.

Designer: Claire Zimmermann; Andrea Davison-Bartolotta
Editor: Jennifer Lombardo

Photo credits: Cover, p. 1 luckyluke74/iStockphoto.com; p. 4 Joe Taylor Cinema/Shutterstock.com; p. 5 Chantal Anders/Shutterstock.com; p. 7 Menno van der Haven/Shutterstock.com; p. 9 VectorMine/Shutterstock.com; p. 11 Armand Rochas/Shutterstock.com; p. 13 John D Sirlin/Shutterstock.com; p. 15 Erica Finstad/Shutterstock.com; p. 17 (bottom) Bjoern Wylezich/Shutterstock.com; p. 17 (top) Jana Janina/Shutterstock.com; p. 18 Home creator/Shutterstock.com; p. 19 Phattaraphum/Shutterstock.com; p. 21 Scott Book/Shutterstock.com.

Printed in the United States of America

CPSIA compliance information: Batch #CWSA25. For Further Information contact Rosen Publishing at 1-800-237-9932.

CONTENTS

Words in the glossary appear in **bold** type the first time they are used in the text.

FLASHES OF LIGHT

Lightning is **visible** electricity. It can be beautiful to see. However, it's also dangerous, or unsafe. It can cause fires, loss of power in homes, and even death.

A flash, or bolt, of lightning is easy to miss since it happens so fast. It often looks like a crack in the sky or an uneven staircase. Some lightning bolts are bright enough to light up the whole sky. Others look more like a camera flash.

Sometimes lightning strikes straight down toward the ground. Other times, it zigzags across the sky.

FUN FACT

IT'S UNUSUAL FOR A PERSON TO BE STRUCK BY LIGHTNING. OF THOSE WHO ARE HIT, 90 PERCENT SURVIVE.

THE SOUND OF LIGHTNING

Thunderstorms form when wet air meets dry air and when warm air meets cold air. The warm air rises and cools. As it cools, water vapor in the air **condenses** and creates a cloud. The cloud keeps rising, and drops of water and ice form inside it. Rain or snow may begin to fall from these clouds.

When lightning strikes, it heats the air and causes it to **expand** quickly. This produces the sound we call thunder.

FUN FACT

BECAUSE LIGHTNING NEEDS WARM AIR TO FORM, IT'S MUCH LESS COMMON DURING SNOWSTORMS THAN RAINSTORMS.

If you count the seconds between when you see lightning and hear thunder, then divide by five, you can tell how far away a storm is.

Seconds	Miles	Kilometers
20	4	6.4
15	3	4.8
10	2	3.2
5	1	1.6

Storm clouds look gray because the water inside them blocks sunlight from passing through. The larger the water drops, the darker the clouds are.

POSITIVE AND NEGATIVE

When something becomes electrified, we say it is charged. There are two kinds of charges: positive and negative. Inside a thundercloud, drops of water, hail, and snow become charged when they move around and run into each other. Lighter drops gain a positive charge. Heavier ones gain a negative charge.

Negative charges gather in the middle of the cloud. Positive charges gather above and below the negatively charged area. When the air can no longer keep the two charges apart, electricity is released in the form of lightning.

FUN FACT

FLORIDA HAS THE MOST LIGHTNING ACTIVITY OF ANY U.S. STATE.

This picture shows the positively (+) and negatively (-) charged areas in a cloud.

LIGHTNING STRIKE!

As a cloud's electrical charges build up, the ground below the cloud also gains a charge. The charge on the ground is the opposite of the charge at the bottom of the cloud. The ground's charge is greater in tall objects, like mountains and trees. These charges attract the cloud's charge. Then, electricity travels between them as lightning!

Water is an excellent **conductor** of electricity. When lightning strikes water, the electrical charge is amplified, or made stronger. This is why you should never go swimming during a thunderstorm.

Some buildings, especially very tall ones, have lightning rods on top. A lightning rod is a specially made metal stick that safely moves the electricity to the ground so it doesn't **damage** the building.

DIFFERENT KINDS OF LIGHTNING

Sometimes you'll see lightning without hearing thunder. This happens when the storm is too far away for the sound to reach your ears. This kind of lightning is called heat lightning.

A blue jet is cone-shaped lightning that glows blue. It happens above a storm's center and then travels up very quickly.

The most common kind of lightning is called intracloud lightning. This happens when lightning jumps to different places inside a cloud, rather than jumping from cloud to cloud or cloud to ground. It's also called sheet lightning.

A sprite is red lightning that happens above storm clouds and can be miles long. Sprites are rare, or uncommon.

FUN FACT

PEOPLE CAN ONLY HEAR THUNDER ABOUT 10 MILES (16 KM) AWAY. HOWEVER, IT'S POSSIBLE TO SEE LIGHTNING UP TO 12 MILES (19.3 KM) AWAY.

FAST AND POWERFUL

Lightning strikes happen fast. They are also very powerful. Each lightning bolt can contain as many as 1 billion **volts** of electricity! It can heat the air around itself to five times hotter than the surface of the sun.

Sometimes there are warning signs that happen just before a lightning strike. People may feel their hair stand up and feel a little tingle in their body. They may also notice a burning smell and a crackling or popping sound. These are all reactions to the increased **static electricity** in the air.

Static is not always a sign of a lightning strike. This boy's hair is standing on end because of the static caused by going down the plastic slide.

FUN FACT

ONE LIGHTNING STRIKE LASTS BETWEEN 0.1 AND 0.01 SECOND.

AFTER A STRIKE

Lightning strikes are very dangerous. They can set houses on fire, make trees explode, and even turn sand into glass!

Many people aren't directly struck by lightning. They are struck by the ground current, or electricity moving along the ground. Although most people survive, they often have serious health problems long afterward. In addition to burns, they may experience memory loss, problems with their lungs and other body parts, or trouble sleeping. The thunder at the site of a strike is so loud, it can seriously hurt a person's ears.

TREE HIT BY LIGHTNING

When sand is heated up, it melts into glass. When this is done by lightning, the formation is called a fulgurite.

SAFETY FIRST

It's not possible to guess where lightning will strike during a storm. For that reason, it's best to be on the safe side. Go inside as soon as you hear thunder or see lightning. Stay inside for at least half an hour after the last time you hear thunder.

If you can't go indoors, put your feet close together and crouch. Keep your body close to the ground without touching the ground more than you must. Stay away from trees and tall buildings.

If you're near water, don't wait until you see lightning. Get away from it as soon as you see that a storm is coming.

ALWAYS LEARNING

Scientists know a lot about lightning, but they still don't know everything. For example, a very rare kind of lightning called ball lightning has been talked about for hundreds of years. However, most people never see it. For a long time, some scientists were not sure it was real.

In 2012, a group of Chinese scientists were able to study a ball lightning event and make a good guess about how it happened: After lightning strikes the ground, it **vaporizes** some of the dirt. This vapor flies upward and reacts with the oxygen in the air, creating a ball of lightning.

MYTH		FACT
Lightning only strikes the tallest thing around. Lie flat on the ground to avoid being struck.	→	Lightning is more likely to strike something tall, but it can strike anywhere. Lying on the ground makes you more likely to be hurt by the ground current.
Heat lightning happens when it's warm out.	→	Heat lightning is misnamed and has nothing to do with how warm the air is.
Lightning never strikes the same place twice.	→	Lightning often strikes the same place multiple times.
If you touch someone who was just struck by lightning, you'll also get electrocuted.	→	The human body does not store electricity. Once the strike is over, a person is safe to touch.
If you wear metal jewelry in a storm, you're more likely to be struck by lightning.	→	Metal does not attract lightning.

Before people fully understood lightning, they came up with many myths, or untrue stories, about it.

GLOSSARY

condense: To change from a gas into a liquid.

conductor: Matter through which electricity flows easily.

damage: Harm.

expand: To get larger and looser.

static electricity: The buildup of an unmoving electrical charge caused by an imbalance between positive and negative charges.

vaporize: To turn to gas.

visible: Able to be seen.

volt: A unit of measurement for the force of an electrical current.

FOR MORE INFORMATION

Books

Bender, Douglas. *Lightning*. New York, NY: Crabtree Publishing, 2022.

Peterson, Megan Cooley. *Why Does Lightning Strike?* Minneapolis, MN: Jump! Inc., 2024.

Websites

BrainPOP: Thunderstorms
www.brainpop.com/science/weather/thunderstorms
Watch a video, play games, and take a quiz to test your knowledge of lightning and thunder.

Kiddle: Lightning Facts for Kids
kids.kiddle.co/Lightning
Read more about what lightning is and how scientists learned about it.

Publisher's note to educators and parents: Our editors have carefully reviewed these websites to ensure that they are suitable for students. Many websites change frequently, however, and we cannot guarantee that a site's future contents will continue to meet our high standards of quality and educational value. Be advised that students should be closely supervised whenever they access the internet.

INDEX